Amazon Echo

Full 2017 Amazon Echo User Manual—Learn Everything You Need to Know About Your Echo from Beginner to Expert

Michael Harris

Table of Contents

Introduction

The Amazon Echo is a great device that you can purchase to help control and manage many of activities that go on in your home and in the office. While many people assume that it is there just to play some music and answer questions, there are many aspects of your life that can be controlled and improved with the help of the Amazon Echo. This guidebook will take some time to talk about the Echo and all of the great tasks that it can help you accomplish.

We will start off by talking about some of the benefits of having the Amazon Echo and why you should consider having it in your home. Then we will move on to talking about how to voice-train the device so that the Echo can recognize some of your speech patterns and you will get back more accurate results than ever before. With the help of the IFTTT app (which we will detail how to download, as well), you can add many of the skills and apps to the Amazon Echo to carry out various tasks around the house or office.

The remainder of the book will discuss all of the cool things that you can do with the Echo and how to get them to work on the device. For example, we will discuss some of the ways that the Echo can help out around the home, how it assists at making bedtime a little bit easier, how to use it at the office to make yourself more productive, and even how to use it along with your workout to try something new and to ensure that you stay motivated. These are just a few of the tasks that you can give to the Echo and the list will continue to grow, but there is sure to be a way for everyone to enjoy the features that come with the Echo.

With the help of this guidebook and the setup tips that are inside, you are going to be able to learn many of the tricks and neat features that come with the Amazon Echo. This is a wonderful tool that can help you out in any aspect of your life that you choose. Make sure to read through this guidebook and learn how the Amazon Echo can work as your personal assistant and make life easier and better in no time.

Chapter 1
Your Introduction to the Amazon Echo

When it comes to working with Amazon, there are many great products that are easy for you to use and enjoy. They started out with the Kindle, one of the first eReader products that took the world by storm and opened up a new industry for the eBooks that are so common now. Since that time, they have released several other products that have been first in their line, including newer versions of the Kindle, their own tablet, the Amazon Fire Stick to help with watching television shows and movies, and the Amazon Echo.

The Amazon Echo, at its most basic level, is a smart speaker that you can control with your voice. This device has so many great things that you can do with it, including controlling some of your smart devices, answering your questions, playing music, sending emails, and ordering food, and there are many other smart features to help you get things done during the day. This device was originally announced in November of 2014 but, when it was released at this time, it was only available to customers who used Amazon Prime. It wasn't until the summer of 2015 that this device became available to the general to purchase.

Once you get the device set up, it is designed to constantly listen for what is known as the wake word. You will use the wake word before your command to turn your Echo on and get it to respond in the proper way. For your Echo, the wake word is set up as "Alexa" by default, but you can change it to "Amazon" or "Echo" if you like the sound of those better or you have more than one device and don't want them to get confused with each other.

It may seem as if "Alexa" is a random word for this device to use, but the Echo is powered by the Alexa app. That is the voice recognition service that helps make the Echo work because it will start to recognize your voice and some of the commands that you give it. Over time, the Echo will store some of the commands that you give and can use its knowledge to better understand what you ask for, thus speeding up the process.

In order to get the Echo to work properly with the Alexa app, you need to make sure that the device is hooked up with the Internet in your home so that it can connect to the cloud to store information about your commands and so that it can automatically get the updates that it needs. The nice thing about this is that Alexa is always learning while it works with you and the more often you use this device, the easier it becomes to recognize your speech patterns and the preferences that you will use when asking for commands, making it really easy to use.

Features of the Amazon Echo

There are many features that you will enjoy using with the Amazon Echo, which is why it has become such a popular product. The Echo is a cylindrical speaker that is a bit over nine inches tall and comes with a remote control, even though you can just use your voice to give the device the commands that you want to use. There are seven speakers that make it easier for the Echo to catch your voice no matter where you are in the room.

Unlike some of the other wireless speakers available, Echo doesn't have a battery, so you must keep it plugged in to get it to work, whether you are using it at home or at the office.

This does make it hard to use if you are on the run but, if you can keep it plugged in wherever you are going, it will work well. Your Echo also needs to be able to connect to the Internet, either through your Wi-Fi or through another secure connection so that it can recognize your voice.

While you can bring the Echo with you to the office and back home if you would like, you should realize that it was designed to be used in one location and not be taken all over the place. It works the best for you to leave it in one place so that you know the best place to talk to get it to respond the way that you would like. Leaving it in one place in the home or in the office helps to ensure that it can hook up to the Internet and reach its cloud-based functions and will help it to work the beset way for you.

You will notice that the Echo is turned on most of the time by default. It is always sitting there waiting to hear your wake word, so the device is on and searching at all times. If you or another person says this word, the device will wake up and start listening for the command or the question that you want it to hear. After the command is heard, it will stream the information to the cloud so that the Alexa voice system can recognize the question and then it will respond the right way. You can also choose to work with the voice-activated remote.

If you ever want to turn off the device so that it won't respond to the wake word and is no longer on, you can click on the Mute button. Then, when you are ready to use this device again, you would just click on the Mute button again and it will turn back on. Some people worry about the device being turned on all the time and recording their conversations, but remember that the Echo is not going to start listening until it hears the wake word and then it only records the questions and commands that you say for learning purposes, placing them on a secure cloud that others aren't able to get to. This device is completely safe to work with and you won't have to worry about it getting personal conversations and information.

To make things easier, the Echo is set up to hook up with your Internet connection at home; the Alexa app is also compatible for many different devices, such as the fire OS, the iOS, and the Android devices if you would like to Echo to hook up with your smartphone, but it is also easy to access through your web browser. It does not take too much time to get the Echo set up with your Wi-Fi, and you will just need to go to the Alexa app and pick which Internet connection that you would like to hook up to and enter the password to make it connect the right way.

Next is the action button, which is right on the top of your device. This is used to help you set up the Echo when you move the device to a new location. Also notice that the Echo does not with a volume button that goes up or down. Instead, the top of the device can rotate to help you to turn it up or down so you can make changes to this based on how well it can hear you. If you are worried about learning how to use some of the other buttons, you should remember that you are always able to use the remote that comes with the device to get this to work instead of using the voice or figuring out which one works on the device. If you get it all set up with your Amazon Echo app, you can use your smartphone as well. There is a way to use this device no matter who you are or which method you feel is better.

As you can see, there are a few parts that you will need to learn about using this device, such as learning which buttons to use and taking the time for the device to learn how to recognize your speech patterns. But once it is hooked up to the Internet and you have used it a few times, you are going to love all the features and the different tasks that you can do with the Amazon Echo and it will become your new best friend when it is time to get things done each day.

Chapter 2
How to Get the Echo Set Up

Now that we have spent some time talking about the Amazon Echo and how easy it is to use, it is time to learn how to get the Echo set up so that you can use it. When you first receive the device, it is just a cylinder that has a few speakers. But once you get it all set up, there are countless things that you can use the Echo for to make your life easier. Here we are going to take the time to set up the Amazon Echo so that you can get it running and then start using it in any manner that helps you out.

Before setting the device up

When the Amazon Echo arrives at your house, it is time to open up the box and make sure that all of the accessories, as well as the device, are inside the box. You should not only find the Echo device, but you should also find the startup guide, which contains some of the commands that you can start out with, and a power adapter. Make sure that everything is there and that nothing looks broken before going further.

After you take all of the items out, plug the power adapter into the Echo device and then into a power supply. Since this device doesn't come with a battery, you will need to always have it hooked up to the power adapter. The good news is that this tool will be easy to tuck into the bottom of the Echo if you need to take the device somewhere with you, but it is pretty easy to use overall.

First steps to setting up after plugging it in

After you have plugged the Echo into its power supply, the Alexa app will say "Hello, your Amazon Echo is ready for setup." You should also notice that a little blue light turns on at the top of this device. This blue light will tell you that the Echo is turning on and getting started. As we work on this setup process, the light may turn on to an orange color; this is not a bad thing, it just means that it is trying to hook up with the Wi-Fi that is in your home. Some of the other lights that you should know about on the Echo include:

- A solid red light: This means that you turned the microphones off. You just need to click on the microphone button to get them working again.
- A solid blue light that has a spinning cyan light: This tells you that the echo is getting started up.
- A solid blue light that has cyan is in your direction when speaking: This means that the Echo has heard your command and is processing what you said.
- All the lights are turned off: This means that the Echo is active and it is waiting to hear what your request is.
- White light: This shows that you are making changes to the volume on the Echo.
- Orange light spinning: This means that the Echo is trying to find where the Wi-Fi network is and is trying to get connected to it.
- Violet light: This means that the Echo has some problems trying to hook up to the Wi-Fi connection.

After connecting

After you have hooked up the Echo to the power source, just leave it alone for a minute. It is time to install a few apps. We want to get the Echo app working and the method that you use will vary based on the browser that you are using or the smartphone operating system. You can look through any of your smartphone stores to find the Amazon Echo app and download it. You can also visit the Amazon Echo page to get this onto your smartphone or your personal computer.

In the rest of the book, we will set up the Echo device to get it to what you would like to use. Once we are done with this setup, you can not only access the app through the remote and through your voice commands, you will also be able to communicate with it through the app. You can also get onto your Amazon account to access the web app for the Echo; this is one of the best methods to use for this because it makes it easier to control and access the Echo device even when you aren't at home and can't talk to it directly.

Once you have set up the app on your smart device, it is time to launch it and use the credentials for your Amazon account so that you can sign in to this app whenever you are ready. At this point, it is time to work on setting it all up through the app.

After opening the app, you need to wait for the orange lights to turn on. You can then connect the smart device right to the Echo, just as you would to a Wi-Fi router. The rest of the process is pretty simple. Inside the app, you just need to follow the prompts that come up. When you are done setting up Alexa, you will hear it say, "You've connected to Echo, go ahead and finish setting up in your Echo app." You will see a screen that says that you are connected to the Echo. At this point, your smartphone should be connected to the Echo as long as you have the app hooked up to the phone and you followed all the prompts that it gave you.

These steps make it easier to hook up the Echo to your Wi-Fi. Remember that you need to have the Echo device set up with the Wi-Fi so that it can work with the Alexa app and it can answer the questions that you ask and the commands that you send, and to get familiar with your voice and speech patterns. At this point, you will still need to be on the Echo app and to follow the prompts from the app. You will basically need to find the network that you would like to use and input the password so that the device can hook up. If you were successful, the app will display a screen that says "preparing your Echo."

After you have had a chance to connect the Echo to the Wi-Fi, and finish going through the prompts so that you can get the last part of the preparing your Echo steps. Now it is time to work on activating the remote so that you can use this with the device if you don't want to use your voice all the time.

Getting your remote set up

One of the accessories that you can use with the Amazon Echo is the remote. The Echo is designed to work with your voice to get your questions and requests, but sometimes it will have some trouble understanding your voice patterns, or you may just not like using the speaking function, and working with the remote can make things a little bit easier.

The remote for the Echo is voice-enabled, just like the Echo itself, and it is battery-powered and wireless. It even has a track pad that helps you to control the audio playback on the Echo quickly and easily. The remote was originally sold separately, but it is now possible to get it with the Echo when you order. When it is time to pair these two together, you can only pair one remote with the Echo, so try not to lose this because it can be a pain to try and get it all fixed. When you are using the remote for the device, you will need to use your wake word before giving a command. Your remote is always going to be listening to what you are saying so that you can give it commands at any time. There are a few buttons that you should learn how to use when it comes to operating the remote:

- Talk: You hold this button down until you hear some tone. Then you can speak to your remote through the microphone. When you are done, release the button.

- Play: This one helps you to get a book or some music to start playing for you.

- Volume up and down: This allows you to turn the volume up or down.

- Next: When listening to music, there are times when you will need to skip to another part of the CD or the list that you are playing.

When you activate the remote, you need to make sure that you peel off the film, then place the batteries inside so that it can turn on. Then it is time to go onto the Alexa app. You can open the navigation menu on the left and select the "Settings" option. Select the device that you are using from the screen and then choose "Pair Remote." If you see something that says "Forget remote," it means that you already have a remote that is paired with this Echo device. If you are trying to replace an older remote, you need to select the "Forget Remote" option so that it releases the older one and allows you to pair this new remote to the device.

When you are ready to pair the remote, press down on the play and pause button and hold it in for five seconds before releasing. The Echo will then search for the remote; it takes about a minute or less to connect. When the Alexa app discovers the remote, it will let you know that the remote is paired. Now you are free to explore a bit about the world of Alexa and learn how to make the Echo device work the way that you would like.

At this point, the app will navigate you to the home screen for the Echo and provide a few tips to help you get started with this device.

Chapter 3
Learning About the World of Alexa

Now that your Amazon Echo is all set up and ready to go, it is time to learn more about the world of the Alexa and some of the different things that you can do with it. We will discuss how to use the Alexa app with your smartphone (which is one of the most common ways to interact with the Echo for most people), and so much more. Let's get started and learn a bit more about how Alexa works.

Using the Alexa app on your smartphone

For the most part, you will use the Alexa app to update the settings in your Echo. Some of the settings can be spoken right into the device or you can access some of them through the Alexa app. Here we are going to use the Alexa app by visiting the home screen. On the home screen, we will see a timeline of all the recent activities you have done with the Alexa app. For example, "Cards" will show a full description of all the commands and requests that you have used. It can also show some features, such as being able to search for the content in the app, some web links to get more information, the ability to show some feedback on your interactions with the app, and even to delete the card.

There are many things that you will be able to do on your home screen in the Alexa app. Some of the options include:

- Change the settings: You can change the settings that will work for the Echo based, on your needs. Take some time to look through these and see what works the best for your uses.

- Access alarms: You can select what alarms you want from the navigation panel on the left. You can decide if you want to turn an alarm on or off.

- Manage the lists: You can go to the navigation panel again and then select on the to-do list or the shopping list that you would like to view, or you can make a brand-new one.

- Canceling a timer: If you want to set a timer, you just need to select the timer from the navigation panel. Then select the option to cancel when you are done with the timer.

- Listen to audio books, music, do a search and more: You can select one of your audio services from the navigation panel. Search for content and then select "Now Playing" to get a look at your queue or the albums as well as to see a history of your tracks. You can play whatever you want from here.

- View your recent activity: Sometimes it is nice to get a look at the activity that has been done on your app. To do this, just select "Home" and it should be listed for you.

Interacting with Alexa

Now that the Alexa app has started and the Echo is ready to go, it is time to look at how you interact with this device to get it to work. There are several methods that you can use to access the Echo device and issue your commands. For the most part, people like to use the voice recognition feature that comes with the Echo, but you can also use the app or the remote that comes with the device if that is easier for you.

Basically, to get this device to work for you, you just need to speak the commands that you want to use. There are many different commands because of all the different apps that work with the Echo, but practicing some of the basic ones can help you to get familiar with using this device and it also gives the Echo a chance to get used to your voice and learn to recognize what you have to say. Some of the basic commands that you can use (remembering to place the wake word that you are using in front of them) include:

- Alexa, softer
- Alexa, mute
- Alexa, volume 6
- Alexa, stop the music
- Alexa, buy this album
- Alexa, who is this artist?
- Alexa, what song is this?
- Alexa, repeat.
- Alexa, previous
- Alexa, connect/disconnect by phone or tablet

- Alexa, pause
- Alexa play
- Alexa, what's popular from (insert band or artist here).
- Alexa, what time is my alarm set for?
- Alexa, set the timer for (pick the time)
- Alexa, snooze
- Alexa, set alarm for (pick the time)
- Alexa, what is the weather?
- Alexa, who is your friend?
- Alexa, what religion are you?
- Alexa, tell me a joke.
- Alexa, see you later, alligator
- Alexa, how much do you weigh?
- Alexa, roll the dice
- Alexa, can I tell you a secret?

- Alexa, what makes you happy?
- Alexa, when is the alarm set for?
- Alexa how is traffic?
- Alexa, is it going to rain tomorrow?
- Alexa, what is the weather in (pick area of choice)
- Alexa, what is in the news?
- Alexa, what is my Flash Briefing?
- Alexa, how far is the sun?
- Alexa, how far is it from here to Canada?
- Alexa, send that to my tablet?
- Alexa, which profile is this?
- Alexa, add (something) to my shopping list.
- Alexa, put my assignment on my to-do list.
- Alexa, play a station from Pandora (or from one of the other stations that you use).

- Alexa, what is the score of (insert game here)?
- Alexa, re-order the tissue paper.
- Alexa, what is the definition of (word of choice)?

These are just a few of the commands that you can send to your Echo. There are countless more and you can find many others that you might like to use. When you are first setting up the Echo, it is a good idea to try out quite a few of these so that the device can get used to your voice and speech patterns. Don't be afraid to try out another question that you have. Since Alexa is a cloud-based service, it can perform a lot of different tasks and can answer many of the questions that you have, so go ahead and ask them.

Using voice training on your Echo

After you get the Echo device all set up, there may be times when Alexa doesn't really understand you or isn't catching the commands that you give. Luckily, this is something that you can work on so that the device is learning your speech patterns. One thing to keep in mind is that Alexa stores your commands to learn these speech patterns, but if you delete them all the time because you are worried about security, your device will have a harder time recognizing your speech patterns and working for you.

In the beginning, Alexa needs some practice working with you. Each person has a different pattern of speech, even if they are speaking the same language, so you will need to use voice training to help the Echo get used to understanding your commands. This won't be hard to do, you just need to have a bit of time and to use some of the questions that are shown above or questions of your own.

To get the most out of your voice training session, open up the Alexa App on your computer or on your smart device. Once the app is open, select the menu that is on the upper left of the home screen. The drop down menu should give you the option for voice training. If you have more than one Echo device, you will need to select which one you are doing the training session on, but the different devices can use the same account, so it is necessary to only use one of them to do the voice training, rather than having to do this over again.

After you select the device that you want to work with, you just need to click to get the session started. The light on the top of the Echo will light up when it is ready for you to get started. You should see a sentence come up on the screen of your smart device and then read it out as you normally would. Click on “Next” when you are done with that one and keep going through. You should have to do about 25 of these statements and you can go back through them again if you want to make sure it all sticks or if you feel that you didn’t say the sentences very well.

Alexa is easy software to work with, you just need to get it all set up and ensure that it is can recognize your voice and what you are saying. Practicing some of the questions shown above or doing the voice training that is available through Alexa ensures that you are going to get the best experience with the Amazon Echo.

Chapter 4
Getting the IFTTT Channel Ready for the Echo Device

Many people have the Amazon Echo wonder if they can hook it up to the IFTTT channel. IFTTT stands for "If This Then That" and it is a recipe that is used by various programs to trigger an example. For example, you could use this to set a keyword in the feed channel to track a topic that you want or even for sending out emails. There are a lot of things that you can use the IFTTT for with the help of the Alexa Channel.

To get the IFTTT channel set up on your Echo, we are going to look at an example of using this app. For this one, we are going to set it up so that the IFTTT can automatically add things to a shopping list that you are creating and how it will appear inside o the app for iOS reminders. Here are some of the steps that you need to do to get this process done.

Step 1:

First, you have to head over to the Alexa channel page. When you are on this page, you can click to connect and, if you don't already have an IFTTT account, you should take the time to get one set up. Once it is created, you need to allow Amazon to share your information with the IFTTT account. After you authorize this, the platform will send you back to the IFTTT.

Step 2:

Next you need to go over to the channel for iOS reminders. Once you are there, just click to connect again. You should remember that, when you continue from this step, you will have to download the iOS app for the IFTTT. After you download that, you just need to follow all of the instructions that come up and then connect it back to the channel. Once it is connected, you can click "Done."

Step 3:

You will need this step to create your recipe. You can do this by clicking on the "This." Navigate to Amazon Alexa to make this the trigger channel, so you need to go to the page that says "Choose a Trigger." Go to Amazon Alexa to set this as your trigger location.

Step 4:

For this one, you have to pick out the action channel that you want to use. Click on the action channel and then scroll down so that you can find the channel for iOS reminders. Once you are here, select the action that says "Add reminder to list."

Step 5:

You need to change the reminder settings to suit your needs. For example, you can add the things that you want to the list and then list them in order of the priority level that you want. When you are done with making changes to the settings, just click on the "Create action" button and finish the process by clicking right on the "Create recipe" option.

The nice thing about working with the IFTTT app is that it has many recipes that you can use with the Amazon Echo. These two apps can work well with each other and many people who purchase the Amazon Echo will just add the IFTT app on to their Echo to make things easier. Since there are a lot of Echo-driven recipes on IFTT, it is easy to create some new recipes to work with your device. You can make as many of these recipes as you would like inside the

Echo, so take the time to practice and try out a few of the recipes.

If you are new to using the IFTTT app or the Echo, you may not be familiar with all of the things that you can do when these two combine and all the great features that come with them. After you have set up Alexa and you get your IFTTT account, take some time to learn all of the things that you can do with the IFTTT app to really utilize the Alexa app to its fullest.

Chapter 5
Using the Echo in Your Home

There is so much that you can to do with the Echo when you first bring it into your home. You can use it to play music and listen to books while you get stuff done in the home and ask any questions that you come up with. You can allow it to control various aspects of your home, such as the lights, garage doors, and more; you can take it to work to help you to get emails sent, ask questions, order an Uber, and even order food. Some people take this a step further and use it for their bedtime routine, for working out, and so much more.

Here we are going to spend some time talking about how you can use the Echo to control your smart home. This means that if you have a smart device somewhere in your home, you can take the control away from your smart device and give it to the Echo. With just the sound of your voice, you can change the temperature setting or turn off the lights, and you can even put all of these on a timer so that they work how you want the second you walk into the door.

Some people like to use these options when they are home to control the things around them without having to worry about where they are, such as if you would like to use the Echo to turn off downstairs lights when you are upstairs and about to fall asleep. Some people like to control what goes on in their home, even when they aren't there. And for those who want a warm house in the winter and a cool house in the summer when they get home, but don't want to pay a bill to keep this going all day, the Echo can handle this to give you the best of both worlds.

There are so many things that the Echo can do in your home to make things easier. Some of those things:

- Locking and unlocking doors. You will need to have the Smart Danalock recipes so that you can auto-unlock or lock the door at certain times, making it easier to keep the door locked at night. You can use the Echo, Tap, or Dot to control this feature.

- Use the D-Link Smart Plugs to add some of your other smart devices to the Echo. You can add smart devices to your home at any time and can easily add them to the Echo to make commands easier.

- GE Appliances Refrigerator Channel—You can even control your fridge from on the road. This channel will allow you to do a few different things. You can blink the fridge door light if it is left open for too long or you can have the Echo set the fridge so that it is on Sabbath mode.

- The Weather Channel—This is really neat if you are always on the run during the morning. You can use it to set your coffeemaker to turn on at a certain time, or at sunrise, or you can turn on the Smart Plug at sunset. This app will also allow you to turn on the air conditioning if the humidity or temperature rises in your home.

- Netatmo Welcome—This app will help you to identify visitors that come to your home and you can change the behavior for each one. For example, if a certain person shows up, you can switch the Smart plug either on or off, and if someone you don't know shows up you can turn on the D-Link of the Smart Plug.

When it comes to making your home a smart home, you can see that there are many different things that you can do with the Amazon Echo. You can use the Alexa Skills as well as the IFTTT recipes. Also, you can choose how many of the neat features you would like to use on your Echo, so don't worry about having too much or even just experimenting with what these have to offer.

Connecting your devices to the Amazon Echo

Before you can hook up any device to your Echo, you need to make sure that you already have smart devices in your home. Regular light bulbs and thermostats are not going to work with this kind of technology, but if you already have some of the smart appliances in your home, you can move the control over from your smart phone or tablet to the Amazon Echo.

Smart home technology has expanded like crazy in the past few years. This means that you will have the power to control many parts of your home with these smart devices and, when you hook them with the Echo, you can control them with just your voice. You can turn your lights on and off, control the garage door, decide when the locks on your doors will lock up or open, set timers to control some of these important aspects, and even have some control over your appliances, such as a smart refrigerator or slow cooker.

If you want to be able to start controlling your smart home with the Echo device, you must decide which type of devices you want to connect to the Alexa. Some of the best options include:

- Lighting and fans—Haiku Wi-Fi Ceiling fans
- Outlets and switches—TP-Link with smart plug with energy monitoring, D-Link Wi-Fi.
- Thermostats—Sensi Wi-Fi Programmable Thermostat, Ecobee3 Smarter Wi-Fi Thermostat, and Nest Learning Thermostat
- Locks—Danalock and Garageio
- Car Control—automatic

The list of devices that work with the Amazon Echo is always growing, so you are sure to find more options that are easier to use as well as some more options of what you can control in your home. You can pick to just try out one of these and see how easy it is with the Echo, or you may be ready to try out all of your smart home features and really see how easy the Echo device can make your life.

An example of setting up a smart home device

You can hook up any of the options that you want in your home to the Alexa so that you can control them from another location. Here we are going to use the Nest Thermometer to get started, but you can follow a similar process no matter what you want to connect to Alexa. We are also going to use the IFTTT option, which should be downloaded with Alexa by now, to help make this easier. In order to connect the Nest channel through the IFTTT interface so it can be used with Alexa, follow these steps:

- Open up the Alexa channel with your computer or smartphone.
- Scroll down and choose the "Smart Home" option
- You will then come to the Device Links tab and from here you can select "Nest" and click "Continue."
- Use your Nest ID and password to log in.

- Now you will see “Discover Devices.” If your Nest account is on your local Wi-Fi network, your Echo will find it.

At this point, you can tell Alexa what you would like to do with the Nest Thermometer and it will make the changes for you. For example, if you are feeling cold, you can tell the thermostat to turn down the air conditioner or turn up the heat, or you can tell it what temperature you would like the house to be.

Working with the SIGNUL Beacon Channel

Another option that you can use is the SIGNUL Beacon Channel. This allows you to set the time you want specific events to occur in your home. There are times you would like to control the locks or the temperature or even the lights, even if you are not at home to tell Alexa.

Think of how nice it will be to get home after a long day and have the temperature set at the perfect place, the doors unlocked for you, and some of the lights turned on, without having these features running all day and increasing your electricity bill. All of this is possible if you find the SIGNUL

Beacon channel in the Alexa skills and get it connected to your Echo.

Before this can work, you will need to set the entry and exit events for the channel. This helps the channel to know when you want the events to begin. The channel can also check out the physical presence of and can streamline the process to help you out. you can pick out which processes you would like the channel to handle, such as just turning on the heat in the winter, or you can have a list of options so that everything turns on for you.

Grouping your lights together

One thing that you can do to make things a bit easier is learn how to group some of your lights together. This can make it easier to turn all of them on and off when you need. For example, if you are upstairs at night and you want to turn off all of the lights downstairs with the Echo, you can group together all of the downstairs lights so you can have them turned off with one command.

If you want to work with this one, you need to make sure that all of the lights are on Wi-Fi; if they are smart lights, this should not be an issue. There are lights and switches that you can purchase from Amazon if you want to keep things easier, but it is also possible to find options from other companies if you would rather use those.

Some of the easiest light bulbs to use include the GE Link and the Wink options, since they are fantastic for helping you to group as well as control all the lights that are inside your home. but first we need to make sure that these light bulbs are connected to the Echo device to help it out. To hook up these light bulbs to the Echo the right way, you will need to do the following steps:

- Open up the Echo app and click on your settings.
- Find "Connected Home."
- Add Wink. When you add Wink, you can see which devices are connected.
- Add these to your group.

You may want to consider hooking up the WeMo app as well; you can find it through Google Play and iOS. This will work with your smartphone and can be hooked up with any of the devices in your home that you want. This can help you to set the lights on timers. So, if you would like to turn on the lights outside at night or have your coffee pot start working when you wake up, for example, this program would help you to do that. It works well not only with the lights, but with all the other smart appliances in your home, as well.

Locking and unlocking your doors

Keeping your home safe is important for your safety. Sometimes it is hard to remember to lock the doors when you are running around and getting the family out the door in the morning and it is a pain to try and find the right key when you get home late at night.

If you make some changes to the locking system in your home, you can let Alexa help you out with this. Once the Echo is attached to the smart locks in your home, you can tell Alexa to lock or unlock the doors whenever you want. This can be helpful t make sure that the doors are locked when you leave the house in the morning or when you need to get in at night. And, if you happen to get in bed and can't remember whether you closed and locked the doors at night, you can just talk to Alexa and give a command to lock the doors. It is that simple.

In some cases, you may be able to get the doors to lock and unlock at certain times during the day on a timer. This can make things easier, as you won't have to have Alexa hooked up to give the command. For example, you could set it up for the home to be unlocked when you are scheduled to be home so that you won't have to fumble for keys or have the doors automatically lock at a certain time of night when you are usually asleep. Don't worry, you can always command Alexa to unlock the doors if you need to get back out again.

Ordering food

You can feed your family with the help of the Echo, as well. Not only can you use it to create a to-do list or a shopping list, but you can learn where some of the best restaurants are in your area. This can be helpful any time that you want to try out something new or if you are in a new area. You can ask Alexa for some suggestions as well as directions to get there.

Another option is to actually order food to have it delivered to your home. There are several options that are available for you and the list is always growing, so your favorites should be available. There are a few steps that you will need to go through to do this. The first is to go onto the Alexa app and find the skill that links to the restaurant you want. You can then either sign up for a new account with that restaurant or link to your account.

Before you can order, go to the account that you have at the restaurant and set up an order. Make this the order that you and your family will normally have from this restaurant and save it. Also set up the payment method that you would like to use.

Once this is set up, all you will need to do is tell Alexa that you want to place an order with that restaurant and it will take care of the rest. You can get the payment taken care of for you since the credit card is on the account and it won't take log for your order to arrive at your door.

Listening to music and your audio books

One function of the Amazon Echo that people like is that they can listen to their music or audio books when they are busy getting chores and other tasks done around the house. First, we will take a look at some of the steps that you will need to do to read your books with the Echo.

You can just tell Alexa to start reading a book of your choosing (just naming the title will be enough). Going through the Alexa app, you can hook it up to your Kindle account. Then, if there are any books listed on your Kindle account, you can say the name or the author of the book and your Echo will start to read it for you. Go forward to a new part, go backwards, repeat, or do something else to customize the way that you hear your eBooks. Audible is a great app to use, as well, since these are already going to read them.

Many people also like to listen to their music when they are on the Amazon Echo. You can make it easy and just ask for a song or a CD from your Amazon account to be played for you, or there are several other apps that you can use to play music, including Pandora, iHeartRadio, and more. You can be in control of your own music choices, save a lot of time and ensuring that you have something good to listen to all the time.

There are so many things that you can do when it comes to making your Amazon Echo work in your home. It is not just a device that allows you to listen to your music and some books while you are working; you can link it to many of the smart appliances in your home to have more control with less work than ever before. Many people will choose to go with the Amazon Echo just because of all the great things that it can help you out with around the home!

Chapter 6
Setting Up an Easier Bedtime Routine for Your Kids

As we go through this guidebook, you are going to learn a few of the great things that you can do with the Amazon Echo. But parents of young children are going to love the next feature of the Echo. We will turn the Echo into an assistant to help you to work with your kids on a safe bedtime routine that is easier on the whole family. There are many ways that you can use the Amazon Echo to help out with the bedtime routine.

Turn the Echo into a timer

As your child is growing, sleep is very important in helping them to have a happy mindset, to help them grow, and to keep them healthy as possible. It is a good idea to set up a routine so your child has a consistent and regular. As a parent, you get busy with cleaning the house, making meals, or doing other things and your children may not make it to bed when they should. To make sure that you are getting your children to sleep at the right time and to avoid having issues with them missing out on the sleep that you need, you can use the Amazon Echo to create an alarm or a timer that will tell you when your child to get to sleep.

It's really easy to get all of those done with the Echo so that you can get things done at night or spend time with your family without having to check the time every few minutes to make sure they get to bed at the right time. You can use Alexa to set up a timer, or even an alarm, so you know exactly what time you need to put them to sleep. In addition, instead of bugging you all the time with asking when is it time for bed, your children can ask Alexa these questions.

Reading stories at night

After running around all day with work and taking care of your kids, it is sometimes hard to fit in a bedtime story for your kids. You can have the Amazon Echo do the story reading. This can be helpful when you are too tired to read the story or you need to have one child listen to a story while you get the other one to sleep. You can sometimes use it to kick off the story time so you can get yourself ready for bed before coming in and reading the rest of the stories.

The Echo can connect to your audio book accounts (there are a few available but the easiest one for you to use is the Audible version), and it will read any of the books that you want to your child. Make sure that you sign up for an Audible account and then link this as one of the recipes that you want to use (through IFTTT). Then you can use the command to tell Alexa which book you would like your children to hear.

Playing lullabies

While you love spending time and having fun with your kids, spending three hours in their room singing lullabies to get them to sleep is not always the best thing in the world, especially if you would like to get some other things done at night. You can use the Echo to sing some of these lullabies for you. You need to setup the playlist that you would like to use on the PC, and then you can ask Alexa to play this playlist whenever you need. This can be great on the days that you are too busy to do the singing, if you are sick or gone and can't do it, or if your child needs to have that music even after they have fallen asleep.

Dim the lights slowly

A good way to help get your kids all ready for bed and to tell their minds that it is time to fall asleep is to dim the lights with the Echo. You will need to hook up the Echo to your dimmer switch and the light bulbs. Some of the light bulbs that are supported include Philips Hue, Insteon, SmartThings, and Wink.

If you already have the bulbs in place, you can synchronize them to the Echo device by setting them up through the Alexa app. After this is done, you can either tell Alexa to dim the lights or you can set it up so that the lights start dimming at a certain time each night. Add this together with the lullaby list that we just talked about, and you are going to have a strong bedtime routine for your child.

Create some white noise

Even if your children don't like to listen to songs to help them fall asleep, you can consider using the Echo device to play some white noise in the background. White noise is usually just a little bit of static, like what you would hear on a radio station when it isn't working, and it is meant to help block out some of the other noises around you (such as cars driving around or the creaking of the house at night) so your child can fall asleep and get into a deeper sleep.

You can use the Echo to help you do this. Most of the white noise machines are kind of expensive compared to the Echo device, especially if you are using the Echo for other things, and you will just need to find a recording of the white noise that you would like. When it is time for your child to go to bed at night, you just ask Alexa to turn on the white noise (or the name of the CD that has it on there).

Talk to your child using the Echo

Another thing that you can use the Echo for during a nighttime routine is to talk to your child. The Echo has a Simon says kind of feature where you can talk into the remote of the Echo, and then the Echo, even if it is in another room, will repeat the things that you say. This can have a lot of good effects on the way that your child sleeps. For example, if your child is asking for something, you can respond by talking into the remote, and your child can hear your response. While this feature can be really nice at times, if your child is upset, it may be best to go in there yourself; Alexa will still use its own voice, even though it is repeating your words, and to an upset child who wants their parent, this may not get the response that you want.

Your bedtime routine just got a whole lot easier with the help of the Amazon Echo. You just need to pick which of the features you would like to implement in your bedtime routine and then get them to sync with the Amazon Echo device and you are set. It only takes a few minutes of your time to get this set up, but it will make night time much more peaceful and even more organized.

Chapter 7
Using Your Amazon Echo to Help Out at Work

So far we have explained how you can use the Echo to help you get a lot of things done at home. Whether you want to use the Echo to help set up your lights or control your thermostat in the home, or even listen to books and help out with the bedtime routine, there is so much that you are going to be able to do with the help of the Amazon Echo. But, in addition to using this device in your home, you can bring it along to work to make into your own personal assistant. There are many tasks that the Echo is fantastic at helping out with in the office and it won't be long until you decide to get an Echo for your office as well. Some of the tasks that the Echo can help you to do in the workplace include:

Order your office supplies

When you are working in your office, you need to make sure that you have all the supplies in order so that your business runs smoothly. This means having simple supplies such as pens, toners, and sticky notes, as well as some of the more expensive options like printers and computer screens. Your Echo will be able to help you to manage these business supplies.

When you are in a hurry and trying to get the ordering done, you may forget some of the items that you really need. The Echo is designed to help you out with this. You can set up a list of items that you need for the office, and then have it reorder whenever you need something. You can also set up the Echo so that it can recognize the different voices that are talking to it, allowing some people to add items to the list, some to purchase things instantly, and others who aren't allowed to make purchase at all. When you are low on some of your office supplies, you just need to tell Alexa and your Amazon account will take care of the rest.

Works with your office software

The Echo can be linked to some of the office software that you use the most. For example, many people choose to link the Echo to their Google calendar so that they can add appointments and other important information. You just need to link your Alexa to the Google Calendar skill and it is ready to go. You can add dates and appointments to the calendar, ask Google to read back what you need to do for the day, and even add some secondary information if you need help remembering what is going on.

This is just one of the types of office software that you can use along with the Echo device. You can integrate it to work with your contacts in Microsoft Outlook to make it easier to send emails using just your voice. The device can add notes to your profile, similar to the way that you would add a new item to your shopping list. After you add these notes to the list, it is easy to have the Echo read them back to you later on so that you can keep up to date, hear information about a client, and so much more.

Turn this into your secretary

The Echo can be the perfect assistant to helping you get things done around the office. It can help you to follow the news, whether you pick a Twitter or Facebook feed or ask it to read through a newspaper while you are getting other things done. The Echo is good at sorting through your emails, based on the criteria that you set in place. For example, you can look up by the subject or the sender to find the information you want. You can have it send you an alert when you get new emails or when an important one comes in and it will read it out for you. When you send a document to others in the office, you can do so with your voice commands, while also sending reminders of when the work is due and checking to see who has finished their part.

Order an Uber

If your work often requires you to leave the office and go to different parts of town, it can be a hassle to always call a taxi or to drive your own car. The Echo can help you to order an Uber ride any time you want. You can ask Alexa to schedule the Uber ride right away so that you can finish up a few things before it gets there, or you can ask for this ride to be scheduled when you need it later in the day. This can save some hassle with looking up the Uber number and can allow you to have some hands-free working time before the ride shows up.

How the DO Note will save you time.

Do you want to get a shopping list done now, or maybe do part of it now and a little bit of it at a later time? The DO Note may be the thing that you need to help you to keep track of the notes that you are always updating (such as getting supplies or working on a shopping list) you can download as well as connect the DO Note (which is an app by IFTTT so it is easy to connect with Alexa) to Google Calendar, Google Drive, Tumblr, Slack, Pushbullet, Evernote, Gmail, Facebook, Twitter, and to Dropbox.

In order to get the DO Note ready to go, you need to enable this skill with the IFTTT website. You can connect the Google Calendar Channel, so that you can write down events, as well as the Evernote Channel if you would like to make some notes while you are out and about each day.

The DO Note can help you out with many things, and not just the notes that you would like to write. It can be added to a few other apps, such as Blogger, to write on your website, the Nest Thermostat to control the temperature of your office, and so much more. Some of the different apps that you can use with the DO Note include:

- Making or changing some events in your Google Calendar
- Adding to your shopping list with the help of Google Drive or Evernote
- Writing down some of your thoughts using the Slack room

- Writing a personal note on Evernote when you are on the go.
- Create a new to-do list and then add or remove the items as necessary.
- Updating your media status with LinkedIn, Tumblr, Twitter, and Facebook.

The stock exchange

The Gobby apps can open up some new doors when it comes to keeping up with the stock market and its quotes. You can use this app to get Alexa to read out the price of the stocks that you have chosen on the NASDAQ or the NSYE. You can also get a summary of this information. You can set up your own personal portfolio to work with Alexa so that you can ask for this information at any time.

If you are in the stock market, it is important to be able to keep up on the stocks that you are buying or selling, especially if you want to see how these stocks are going to affect your growing business. Having this app on the Alexa can save you some time because you can listen to the information, rather than always having to search for it while getting other things done.

Finding a place to eat

There are many different situations that can come up when you are working in business. One of these may be a time when you will need to take some clients out to eat. If you want to find a place that is brand-new and has something different to offer, you can ask Alexa for some restaurant ideas. You can ask for those that are near your location, ask about a certain theme or type of food, or ask based on another category that you would like. This makes it easier to plan things so that you can do what you do best while also impressing your clients with your great choice of restaurant.

Another time that you may want to use this option is if you are traveling on business to another town. You can ask the Alexa app about some of the restaurants that are in the area if you aren't familiar with it. This makes it easier to find the meals that you will want to eat and ensures that you don't go hungry or that you aren't stuck with going to local fast food restaurants that are right by your hotel.

Setting up your DocSend Channel

This can be a great option to use if you need to keep track of all the documents that you are sending throughout the day. You can connect the different recipes together to determine when someone has received your document and even when they have completed reading it. You can also send a follow-up email after someone has read the document to check that they got the update or ask if they have any questions. Some of the channels that you can use to do these skills include:

- Gmail—This is the most popular one and can make it easy to connect with the DocSend channel. Any time that you have a visitor to the group, you will receive a notification.

- ORBneXt—With this option, you will receive a notification when a visitor comes on or even when someone read through the whole document. This can help you to keep track of whether everyone on the team has had a chance to read through the document and you can remind those who haven't had a chance that they need to do it.

- If Channel—This is another one that allows you to tell when someone is on the document with you and who has had a chance to read through the whole thing.

Sending documents to others on your team with the help of Alexa can make things so much easier. You can interact with each other, ask questions, and keep track of the progress of each other. It is simple to handle and can help to improve communication between you and the other people working with you on a project.

Setting Up Square Channel

This is a great tool to use if you accept online payments. If you work from home or have another business that could do work online, you should set your Echo up to work with the Square Channel. You simply need to go and set up your own Square account before you can activate it on your Echo, but once you do, you can enjoy the many features including:

- Payments—Receive an email any time that money has been sent to your Square account so that you can keep track of which customers have paid you and which ones are still owed.
- Refunds—Any time that someone uses your Square to make a refund, it will add another line to the spreadsheet. You can even receive an email when any of

your accounts receives a refund of a certain amount so that you can keep track of the money you are spending.

- Funds going into an account—Any time that there are funds going to your account, whether it is a new payment or a refund or some other thing, you can receive an email for your account.

You can use Square with a number of other channels including Gmail, ORBneXt, and DocSend. Just make sure that you have the Square Channel set up and that you are sending all of your business transactions through here in the future.

Adding in your blogging content

Do you like to spend time blogging? Is this a way that you help to promote your business across the world, or are you more interested in following a certain blog in the hopes of keeping up to date on the things that matter the most in your line of work? The Amazon Echo can make it easier to add these blogs to your device. You can use one or several channels at a time using channels such as Tumblr, WordPress, and Blogger. You can even use some other options, such as Salesforce, Quip, Slack, or LinkedIn to add your own personal business network.

Picking the blogging network that you want to use for your work life or for your Echo to keep track of can be a challenge. Some people choose to go with many of these, but depending on your business, this could end up being a ton of extras on the Echo. Some of the best blogging channels that you should consider to keep up to date with your work include:

- Tumblr—This is a great site to use if you want to publish a lot of photos. It can work with Flickr and Instagram so you can merge all of your information and post a blog all in one place for your customers to take a look at.

- Blogger—Blogger is great if you want to integrate a ton of other things in with it. You can use it to blog any images you have on Dropbox and even share these new posts on Facebook to easily reach your consumers and let them know when new information is posted. If you are interested in doing Vimeo vids, they are easy to add onto Blogger, as well.

- WordPress—WordPress is a great place to start if you are a beginner with blogging. It is easy to use and really stable and you can make things even better when combining with the Echo. You can use WordPress with Tumblr, add YouTube videos when needed, add your pictures from Instagram, and even publish your blogs online, all thanks to the Echo.

There are other options when it comes to creating a log with the Echo; you simply need to choose the blogging site that works the best for you and then set up your account before sending over the information to your Alexa account. Once you do this, it is easy to tell Alexa what you want to get done with a few simple commands.

Blogging can make a big difference to your business and can help it move forward. With good content, you can bring in more advertisers who will work with you to place content on your page. You can sell different products to the customers who come and visit you and get ranked higher in search engines. Why not spend some time learning how to work with the Alexa software so that you can get your blogs published and running in no time?

Newspapers

When you are in business, you need to make sure that you keep up with the news. The Echo can help you out with that. You just need to pick out some of the newspapers that you would like to keep up with, including the HuffPost, *New York Times*, and more and then link the skills that correspond with these to the Echo. Then, when you get into the office in the morning, you just need to ask Alexa to start reading the newspaper or even just the headlines on that particular newspaper.

Imagine how nice it will be to work on some other things, even checking emails, while Alexa reads the news to you. This can save a lot of time, because reading through all the newspapers on your own can be time-consuming and they don't really leave much time for you to get other things done at the same time. But, with the help of Alexa, you can hear the news and get more done at the same time.

The Echo is a great tool to bring to the office with you. It can help you to get more done while you are working and it ensures that you are being as efficient as possible. Once you bring this device into the office with you, you are going to wonder how you were able to get anything done around the office without it!

Chapter 8
How the Amazon Echo Can Help Increase Your Fitness

Do you want to lose weight, or at least get a better handle on the fitness and workouts that you are doing? Do you find that it is hard to stay motivated and get things done when it comes to working out and meeting some of your goals? The Echo can help you out, not only in your daily life at home and at work, but it can also be a great tool when you want to increase your fitness.

There are many ways that your Echo can help you out. You can add some third-party extensions to the device, using the skills that you can download with your Alexa app, to use the types of exercises that you want.

You can use many other tools to help you stay active and be as fit as possible; nothing is as effective on its own as it can be with the Echo. You can add your exercise equipment and even your Fitbit to the Echo and get even better results than before.

Depending on the type of exercise equipment you are using, you can attach it to the Echo. This can be helpful in several ways. You can use the Echo to change information on the machine, such as the amount of weight or even the incline. If you are uncertain about what to do on the machine (if the machine is new or you want something different for the workout) you can talk to Alexa to learn something new.

You can also use the Alexa feature on your Echo to keep up with health trends. The *New York Times* is a great place to get this information and can be easily linked in with the device. Add this to your Echo and start to get emails sent to you with tips and tricks for the best health and exercise routines.

You can use the fitness part of your Echo to turn on the Withings app to keep track of your body measurements. You simply need to add the Google Drive Channel and the Withings Channel so that you can record the measurements and have them stored. If you like to keep track of your sleep schedule and the amount of steps that you take, it is a good idea to download the Fitbit Channel and have this information sent to Google Drive.

As you can see, there are so many things that you can do with the Echo that can improve how well you keep your health up. You can choose to keep track of your body measurements, sleeping times, and steps, and you can even learn some new tips and tricks to keep you moving throughout the day. These are simple to set up; you simply need to open up IFTTT and look for the recipes that are listed there.

Connect the Echo with Fitbit

Fitbit has quickly become one of the best fitness tools on the market. There are so many options available that can help you keep track of your daily activity. Many people who feel that they are healthy and active are surprised when they put on a Fitbit and find out that they aren't moving that much or aren't burning as many (estimated) calories as they had assumed.

The Fitbit is a great way to track many aspects of your health, including your sleep schedule, how many steps you are taking, how long you are working out, your resting heart rate, and more. For many people, it is a great way to motivate themselves to get up and get moving, making it easier to stay healthy and lose some weight.

You can choose to link your Fitbit with your Amazon Echo to update and keep track of your daily fitness goals. While this is still one of the areas that Amazon needs to work on because the interaction isn't as smooth as with some other programs, the Fit Assist will take a look at your goals and give you some facts and tips about fitness and health.

Alexa is not going to store the data about your daily activities with the Fitbit (you can download the Fitbit app to keep track of this if you wish), but you can use the information that Fit Assist sends you to make new goals each day. Just add this part by going to alexa.amazon.com and looking under skills.

If you would like to connect your Fitbit to the Amazon Echo, you need to go through the IFTTT. Go to the Alexa website and connect both the Google Calendar channel and the Fitbit channel. Make sure that your goals are set on the Fitbit so that Alexa can send you the right reminders.

Once both of these channels are set up and linked with your Echo, Alexa can help you adjust your sleeping schedule and tell you when it is best to go to sleep, and will also provide you with a spreadsheet of your activity through Google sc that you can keep track of everything. This can be a nice way to see your progress and get reminders of how to get enough sleep to lower stress and to keep functioning through the day.

Skills available through Alexa for exercise

For those who are all into the fitness and health, or even for those who are just getting into it and want to make sure they are on the right path to the healthiest life possible, there are a lot of great apps that you can use with your Echo to make things easier. Some of the options that you may want to try out include:

- FitnessLogger—when this is used with Alexa, it will help you to record your specific fitness schedule. It doesn't matter what fitness schedule you are on, or even if you are trying to mix it up, this will help you to record the exercise and then compare workouts throughout the week. You simply need to say "Alexa, ask FitnessLogger for all supported exercises" and then you will receive a list of all your exercises. This helps you to see if you're being consistent or if you need to pick it up a bit.

- 7-minute workout—This will help you to cut out the fat in your life and even to lower stress. You don't need to have an hour of working out each day to see results. Sometimes a few minutes is enough to make a big difference. Or maybe you just had a stressful day and need a few minutes to reduce the stress. Seven minutes of a quick workout with the help of Alexa can make all the difference. All you need to say is "Alexa, start seven-minute workout" and you are ready to go.

- Training tips—If you are new to working out, make sure you set up the training tips. This can help you to learn the best workouts for your needs and to get the most out of your gym time. You just need to ask Alexa for tips for the day.

- Recon Channel—This app is a bit more expensive than the others, but it is top of the line and will boost how well your Echo can take care of your workouts. It is meant to track your fitness by projecting your metrics to an eyepiece. You can see how well you are doing on the workout while moving, without having to stop to check out the statistics. You can also get sports news

DubNaiton updates, calendar updates, and more from here once you get it set up with the Echo.

- ALOP-Pilates-Class-Skill—If you are interested in learning how to do Pilates as a good start to your workout or as a way to get up and stretch a bit more for sore and aching joints, this Pilates app is the best one for you. You can simply add this to your Alexa skills and then tell Alexa that you want to start the Pilates class. This skill will take you through a whole exercise schedule to help you get something new each day. If you would like to set it up, just check out ALotOfPilates.com.

You can also use the app to track your food intake, the exercises that you are doing, and the various measurements that you need to take. This can help you to get started on your new workout program or to make your current one even better. No one wants to go on a plan and find out that it won't be that effective, but, with the help of your Amazon Echo, you can add a few different apps and get the best results in no time.

Using the Amazon Echo is one of the most efficient ways to keep track of your workouts and to ensure that you are going to get the results that you want. There are apps for those who are just starting on their workout adventure and some for those who have been at it for a long time. Use Alexa to link your favorite apps and find how to keep your body as healthy as possible.

Chapter 9
Other Fun Things You Can Do with Alexa

In addition to the Alexa working well as a personal assistant in your home and in the office, there are a lot of great things that you can do with Alexa that are fun and can increase its functionality. In this chapter, we will learn about some of the fun apps that can be added to Alexa to make it even more enjoyable.

Connect to the Ooma Phone System

If you like automation, the Ooma Phone system will offer some comfort and value. Alexa will help with this because it can connect you to the voice-command operations that come with this phone system. You will not be able to make a phone call through this system with Alexa, but it is possible to have Alex send a voice mail for Ooma or for Alexa to place a call to a person or a certain number that you choose.

If you would like to connect the Echo with the Ooma Phone System, there are some simple steps to make this happen. First, you need to open up a free account with the 1500-minute lifetime limit. This will make it possible to get up to 300 minutes of talk time with the Ooma system each month. Then you will need to go through and download the app for Ooma. In order to set up this system, do the following steps:

- Open the Menu icon in the Alexa app. You need to do this on your phone.
- When you are there, click on "Skills," go into the search box, and look for Ooma.
- Once this shows up on the screen, you will see "Enable" on the drop-down list.
- Tap "Enable" and then create an account with Ooma.
- At this point, you will need to enter some information, including your zip code, email address, and phone number.
- You will need to take some time to verify your email and phone number so that the account can be activated.

Find out the price of gas

If you travel around quite a bit, are in a new area, or just want to know if the prices have changed, you may find that the Gas Price Finder app is great to use. You may not see a huge difference in the gas prices within a few miles of your home, or if you travel from one town to another for work, but over time that little bit of change will add up. Some people use this if they live in one town and work in another, because they can find the cheapest gas prices to use.

When using the Gas Price Finder, you can enter the zip code of the place you want to use for finding gas as well as your car number so that you can compare some of the prices where you are going. You can then make travel plans, whether it is in a short distance or across the country, depending on the prices that are presented. In order to get this hooked up with your Echo device, you will need to follow these steps:

- Go to the website for the Amazon Echo and then go to "Menu" and "Settings."
- The tab for "Connected Apps."

- Find the link for Gas Price Finder and click on "Enable" to get it hooked to the app.

Learning some pizza facts

If you are looking for something fun to do while waiting for a meeting or you are looking for something to do with the family, pizza facts can be enjoyable. You can download the Pizza Facts app to learn about pizza to help keep you entertained and distracted until a meeting or you have something else to do. You can enjoy any of the other trivia apps or other fun game apps that you would like to use.

Beer advisor

This is a great app to use if you want to pair the right beer with your food. This one is good for newer cooks as well as some of the older ones. The Beer advisor helps you to find the right beer that to pair with the food you cook. You can download and enable the Beer Advisor app and then just ask Alexa for the advice that you want by saying "Alexa, launch Beer Advisor." You can follow this with the type of food that you are having, such as steak, and Alexa will list the beers that you should drink with it. Now you will not only have a great meal to enjoy, but the right kind of beverage to make it taste even better.

SciGuy

If you like to learn more about science or want to impress others with your knowledge, you need to get the SciGuy app. This will work with the IFTT website as well and it is good for helping you to learn new facts every week. You can even make this work like news updates so that the Echo will read them out to you. You will love these facts and how varied they are each week.

Chemistry Genie app takes this further by giving you updates on the elements in the periodic table. This is a good one for a student who may have an exam coming up or for someone would just like learning more about the various elements. You can get regular updates, as well.

Games

There are many games that you can play with your Amazon Echo; you just need to make sure that you download the right apps to make this work for you. Some of the great games that you can add for lots of entertainment with the Echo including;

- Jeopardy: This app has twelve categories to choose from; you can pick six of them to play with. Once you are through this, you can play your sixth clue in each class or you can bring out the quiz from the previous day if you want to keep going. This is a great way to learn some new things, have some fun, and play a great game.

- Word Master Game: You can play this game in either Portuguese or English. You will need to download the app, but then you can use the app to sharpen your skills in English and learn some new words. The game will start with eight letters and you have to make words that have a minimum of the last three letters while being times. You can go along with Alexa and play together and any time you are stuck on a new word, you can end the game and exit or start over.

- Bartender: This is a great game to play if you are interested in how cocktails are made. You will just need to ask Alexa what is the cocktail (name) and this app will allow you to hear the recipe to make this cocktail, as well as some variations if you would like to try out different ways.

- The Wayne Investigation: This is a good mystery and action game that was designed by DC Comics along with

Warner Bros. It will guides you through an adventure game (through audio) where you investigate the murder of Bruce Wayne's parents. You will be an investigator in Gotham City and can use various commands to navigate the scene. This is a great game to play when you want to try something new, or if you like DC Comics.

These are just some of the games that you can play when it comes to using the Amazon Echo. They are easy to download, most work through the IFTTT app, and you can pick as many of these as you would like. There are also always more games coming out for you to pick from, so make sure that you check back to see if some of your favorites become available for the Echo.

Chapter 10
Some Troubleshooting to Help Out with the Amazon Echo

For the most part, you will run into any issues when you use the Amazon Echo. This is a great device for you to enjoy and, if you followed some of the tips that we gave above, you shouldn't have any issues with getting the Echo to work for you. With that being said, there are some little issues that can come up with the Echo, and most of them are pretty easy to fix. In this chapter, we will look at some of the common issues that may come up with the Amazon Echo and will ensure that you can get the most out of your device from the moment you receive it.

The Echo can't find your devices

One of the benefits of using the Echo is that it can sign on to your smart devices through the home so that you can run them with a simple vocal command. This makes things easier because you don't have to run around the home to get the tasks done. You can set up your locks, your lights, your fridge, and other appliances around the home to work with your Alexa so that you just have to give some simple commands to turn them on, off, and more.

But there are times when Alexa will have trouble finding these smart home devices. If Alexa is not able to find the devices, it is really hard for the software to control them and your commands aren't going to work. The first thing that you should check is whether the device is supported natively in the system. If it's not, you may need to take a few more steps to get it to work with the Echo.

The list of those apps that are natively supported with the Echo is growing; some of the devices that work well with it are Ecobee3, Philips Hue, Nest, lifx, Insteon, Honeywell, and Winx. There are many other devices that are supported by Alexa if you use the Skills. This is not going to be officially recognized by Alexa, though, so you will have to add them in to make them work.

To add these new devices, you need to open the Alexa app and go to "Smart Home." Tap on "Discover Devices" and go to "Your Devices." You can see here whether the devices are natively supported. If they're not, you have to go through the IFTTT channel to add the devices.

If you've added the devices to your Alexa account, either natively or through IFTTT, and Alexa is still not able to find them, there are a few other solutions that you can try, including:

- Check which command you are using with the Echo. This can make a big difference on whether Alexa can understand you or not. Even small changes in the names or the phrasing of your smart devices can confuse Alexa and it won't be able to connect. Look up what the device is called and then use this in your command.
- The issue may be with your smart device rather than with the Alexa. Some have issues with their software that will make it hard for it to stay connected to Alexa. Checking the power cycle of your devices can help you to see if you have any connectivity issues. Sometimes you may need an update to the device to fix some of the software issues.

If none of this is working for you, you should reboot the speaker and remove the device from your Alexa, then add it back to see if that helps.

Alexa has trouble staying connected to the Wi-Fi

At times, your Echo device may have some issues staying connected to the Internet. When this happens, it is impossible to interact with Alexa because it can't record you if it is not online. The first thing that you should do is check your personal Internet connection. If you are having trouble with that, it is time to do some troubleshooting to fix this issue or talk to your Internet provider to see what is up.

If nothing is wrong with your Internet connection, try out a few other things to get the Echo to stay online. Some of the things that you can try include:

- Reboot everything—One of the easiest things to try when your Echo won't connect is to reboot everything. Just turn of the power and give it some time to reset. Unplug the power adapter and perform a power cycle on the router and modem at the same time. Keep the Echo off until the modem and router have time to come back on. You should notice that the Echo is working much better after this time.

- Move your Echo—Sometimes the issue is that the Echo device is just too far away from the router. If you have the router in the basement and the Echo is on the second floor across the house, you may have found your issue. You should have the Echo and the router located centrally in your home. Having them higher up in the room can help avoid issues with barriers. Keep the devices closer together, as well, so the Internet signal doesn't have as far to travel. Try to keep the Echo away from metal objects or the wall by at least 8 inches to help with connections.

- Reset—This is not always the best option to try because it will get rid of some of your settings and it doesn't really tell you what the original problem was. But it can help to solve almost every problem that you have with your device. It will set the device back to the original factory settings so you will have a clear slate. To reset the Echo, follow these steps:

- Get a paper clip or a similar item and locate the reset button is at the bottom of the Echo near the power adaptor.
- Hold this button in until you notice the top of your Echo turns orange.
- Now you have to wait for the light to turn completely off and then turn back on again.
- Now you will need to bring out your computer or smart phone and open up the application for Amazon Alexa.
- Walk through the whole setup process for the Echo again, just as you did in the beginning.
- Once you are done with these steps, the device will be back to its factory settings and ready for you to use again.

- Call customer service—If none of these tips help, it may be an issue with your hardware or the service provider. The easiest thing to do is call your Interest provider and see if a spotty connection is the issue. If that doesn't help, talk to customer service at Amazon. Be aware that they will have you go back through all these steps just to ensure that you have tried everything. But they are the best bet when it comes to figuring out how to solve the issue.

Alexa is having troubles hearing you

Over time, you might notice that Alexa is not hearing you as well as when you first got the program. There may be something wrong with the speakers, but you should try a quick reset to see if it will help.

Turn off the speakers and then turn them on again. Do a test run and see if the Echo can hear you properly now. If this doesn't seem to fix the issue, try moving the object around a bit. You need to make sure that the Echo is a minimum of 8 inches from the wall and that there are not any obstructions that make it hard for the speakers to hear you.

The obstruction can be something very simple. For example, having the air conditioner on can muffle your voice and make it hard for the Alexa software to hear you. You may not notice this if you bought the device in the winter, but once summer comes and the loud air conditioner turns on, it can be hard for Alexa to hear.

Moving the device closer to you, or at least farther away from the object that is blocking it, can make it easier for Alexa.

You may also need to work with the voice training app to get Alexa to understand what you are saying. With this app you will read out 25 phrases, using your typical voice and from a distance you normally would be from the device. This helps Alexa to recognize your voice so that it can respond better.

The Echo won't turn on

You may have trouble with the Echo not turning on. This may happen for a few simple reasons, but it can be frustrating. Here are some of the reasons that the Echo won't turn on:

- Not plugged in—You should to see if the power cord was unplugged or if it is falling out a little bit. If you are using a power adapter, check to see that it's working properly.

- Faulty power cord—If everything is plugged in correctly, check the power cord. See if you can spot any damaged or frayed areas. If you do find this damage, make sure you replace the power cord as soon as possible.

- Faulty hardware—In some cases, you may have some issues with the hardware inside the Echo. If you have tried some of the other options and it is still not working, you may need to call Amazon's customer support to see about getting replacements for the Echo hardware.

These simple tips may make it easy to get your Echo device up and working the way that you would like. Some issues may come up with an Echo, but usually it is something simple that we can fix in no time. Try out a few of these tips and get that Echo device back up and working in no time!

Chapter 11
Easter Eggs on the Echo

We have talked about some of the great things that you can do with the Amazon Echo to help your life be easier. Whether you are looking to make your home into a smart home and control the lights, locks, and more, or you want to bring it into the office, to your workout, to your bedtime routine, or to play some great games, you can get the Echo to work.

But there are some other great things that you can do with the Amazon Echo. Some people like to mess around with the device and find some silly questions or other statements that can be fun with Alexa. Many of these, which are known as Easter Eggs, are unknown by some users, but they can make it even more fun to work with Alexa and may give you a good laugh.

Here we are going to take a look at some of the Easter Eggs that can be found in the Echo. Remember that you can use any of these you would like (and they will all elicit a response from the Echo) but you do need to add the wake word before each of them. Some of the best Easter Eggs that you can use with the Echo include:

1. What is the meaning of life?
2. What is the best tablet?
3. Make me a sandwich.
4. I am your father.
5. What is the loneliest number?
6. What is your quest?
7. Beam me up.
8. Who is the fairest of them all?
9. How much is that doggie in the window?
10. Who ya gonna call?
11. To be or not to be?
12. How tall are you?
13. Can you give me some money?
14. Knock knock?

15. Party on, Wayne.

16. Do you have any brothers or sisters?

17. Tell me a joke.

18. Are you lying?

19. Roll the dice.

20. What is the sound of one hand clapping?

21. Do you like your name?

22. Am I pretty?

23. Do you smoke?

24. Are you thirsty?

25. What are you made of?

26. Who is your best friend?

27. I'm hungry.

28. You're silly

29. I'm sad.

30. You hurt me.

31. Where do babies come from?

32. Say the alphabet.

33. Do you want to build a snowman?

34. May the force be with you.

35. Do you have a partner/boyfriend?

36. Can I ask a question?

37. Tell me something interesting.

38. Are you smart?

39. You're wonderful.

40. How high can you count?

41. What number are you thinking of?

42. Happy holidays.

43. Say a bad word.

44. Can you smell that?

45. One fish, two fish.

46. Will pigs fly?

47. How old are you?

48. Were you sleeping?

49. Do you dream?

50. Are you crazy?

51. Are you real?

52. Can you lie?

53. Do you have a last name?

54. Did you get my email?

55. Do you want to go on a date?

56. Honey, I'm home!

57. Guess what?

58. Make me some coffee

59. Is there life on other planets?

60. Play it again Sam.

61. Say cheese!

62. What are you going to do today?

63. What is love?

64. What do you want to be when you grow up?

65. Will you marry me tomorrow/Why is the sky blue.

66. Random fact.

67. You're wonderful.

68. Why so serious?

69. What's the answer to life, the universe, and everything?

70. What's your birthday?

71. What's your favorite food?

72. Are you a robot?

73. Who stole the cookies from the cookie jar?

74. Do you believe in love at first sight?

75. What color are your eyes?

76. What are you wearing?

77. How much does the earth weigh?

78. What do you think about Cortana?

79. What is a day without sunshine?
80. Where have all the flowers gone?
81. How much wood would a woodchuck chuck if a woodchuck could chuck wood?
82. Do you like green eggs and ham?
83. It's a trap.
84. Winter is coming.
85. Who lives in a pineapple under the sea?
86. High five!
87. It's a bird! It's a plane!
88. What does a cat say?
89. I have a cold
90. Will it snow tomorrow?
91. Why are there so many songs about rainbows?
92. What is my mission?
93. Do you know everything?

94. Do you have a job?

95. Do the dishes?

96. You complete me.

97. Why do you sit there like that?

98. Who is your role model?

99. Twinkle, twinkle, little star.

100. Don't listen to him!

101. What is your favorite ice cream?

102. Can we be friends?

103. Can reindeer fly?

104. Who is the shortest person in the world?

105. What are you thankful for?

106. I don't know.

107. Where's the beef?

108. Entertain me.

109. Can you tell me how to get to Sesame Street?

110. Life is like a box of chocolates.

111. How old am I?

112. Do you want a kiss?

113. Take me to your leader.

114. Live long and prosper.

115. Show me the money.

116. What are the five greatest words in the English language?

117. What is a prime number?

118. Can I tell you a secret?

119. What is happiness?

120. Who is your best friend?

121. What size shoe do you want?

122. Tell me a poem.

123. Do a barrel roll.

124. Say hello to my little friend.

125. Do you know Siri?
126. How many roads must a man walk down?
127. Romeo, Romeo, wherefore art thou, Romeo?
128. How many licks does it take to get to the center of a tootsie pop?
129. Is the tooth fairy real?
130. Where's Waldo?
131. What's your birthday?
132. Where are my keys? (you will need to ask this one twice)
133. Have you ever seen the rain?
134. Speak.
135. Are you my mummy?
136. This statement is false.
137. What are the seven wonders of the world?
138. I'll be back!

139. Say you're sorry.

140. Tell me a tongue twister.

141. Are you in love?

142. Who loves ya, baby?

143. Which came first, the chicken or the egg?

144. Who's going to win the Super Bowl?

These are just a few of the Easter Eggs that you can find in the Amazon Echo. You can ask it pretty much any of the questions that you want, even if they are silly like some of these, or repeat some lines from your favorite movies to see what the Echo will repeat back to you. This is a great way to have some fun with your Echo and it ensures that you will be able do some voice recognition with the device as well. Give a few of the options on this list a try or make up some of your own and have some fun with the Amazon Echo, in addition to putting it to work as your personal assistant as well.

Conclusion

The Amazon Echo is one of the newest products that is available through the Amazon company. It is easy to use and, regardless of how you choose to use it, you will find that it is a good personal assistant for getting things done. There are countless things that you can use the Echo for, whether you bring it to work, use it on your workouts, implement it into your bedtime routine, or just use it at home, and it is soon going to become one of your new favorite tools.

This guidebook talked about the Amazon Echo and many of the functions you can carry out with this device. We started out by learning how to use the Echo and how to get some of the voice commands to work for you. This includes some voice training to make it more effective, so the device is better able to recognize some of your personal voice patterns but, once you are done, the Echo will be able to understand what you have to say.

We hope that you learned some great stuff that you can do with the Amazon Echo, including reading books, controlling your lights, sending out emails, using it in your workouts, and so much more. This guidebook is meant to be a guide to the use of the Echo that will help you get it to work the best for you.

Whether you are thinking about purchasing the Amazon Echo or you already have the device and want to learn more about how this device works and all the cool things that you can do with this device, make sure to check out this guidebook and learn how to make the Amazon Echo work for you!

49317354R00065

Made in the USA
Middletown, DE
13 October 2017